How to fix, r troubleshoot your ~~~ like a professional

A car book for all levels: auto mechanics fundamentals

Alexandre Tesla, Alan Delfin

© Copyright 2018 – Alán Adrián Delfín Cota. All rights reserved.

The contents of this book may not be reproduced, duplicated or transmitted without direct written permission from the author.

Under no circumstances will any legal responsibility or blame be held against the publisher for any reparation, damages, or monetary loss due to the information herein, either directly or indirectly.

Legal Notice:

This book is copyright protected. This is only for personal use. You cannot amend, distribute, sell, use, quote or paraphrase any part of the content within this book without the consent of the author.

Disclaimer Notice:

Please note the information contained within this document is for educational and entertainment purposes only. Every attempt has been made to provide accurate, up to date and complete, reliable information. No warranties of any kind are expressed or implied. Readers acknowledge that the author is not engaging in the rendering of legal, financial, medical or professional advice. The content of this book has been derived from various sources. Please consult a licensed professional before attempting any techniques outlined in this book.

By reading this document, the reader agrees that under no circumstances is the author responsible for any losses, direct or indirect, which are incurred as a result of the use of information contained within this document, including, but not limited to, —errors, omissions, or inaccuracies.

TABLE OF CONTENTS

INTRODUCTION

These days, cars are the ultimate means of transport because they are very comfortable as well as very convenient. Cars are a major part of a fluidly functioning society. We use cars to transport goods, get to and from work, to run errands, respond to emergency situations in a timely manner and to have fun. But cars can be an incredibly costly investment. The cost of a car goes far beyond what it takes to buy it.

Owning a car means paying for gas, general upkeep, registration, new tires or brakes, inspection stickers, major repairs from constant wear and tear or damage caused from a wreck of fender bender. Although these costs may seem overwhelming, there is a way to cut down the price of owning a car significantly. The path towards a longer life for your car with fewer repair costs is by routine maintenance.

By far the greatest cost of owning a car, aside from maybe the original purchase, are repairs. Cars usually break down without any warning and while gas, registration and inspection and even oil changes are most likely a part of your monthly budget, the hundreds needed for major repairs are usually not.

Therefore a car should be treated much in the same way that a pet or even a child is treated as far pre-emptive measures to keep it healthy. Pets and children have routine doctor, vet or dentist visits in order to keep serious conditions at bay. In the same manner, cars should be checked for any sign of possible damage before a major breakdown happens. Failure to maintain can be the cause of serious car accidents.

We all avoid maintenance on one pretext or the other but forget its core advantages, like low recurring expenses and optimum product utilization, especially in the case of car maintenance. By doing regular maintenance on our car, not only we can save money on car repair, but also enjoy more pleasant drives.

Car maintenance is not a big technical job, and a novice with minimal car-related knowledge can also do it easily. You can initially start with tires. Tires are among the most expensive, regularly replaceable and neglected items on the car.

One must follow some basic rules for tire maintenance.The first and the foremost are checking tire tread for abnormal wear patterns. On finding any symptoms of abnormal wear in specific areas, tire alignment needs to be done to prevent further deterioration. Along with it, you must check tire pressures once a week, for longer tire life. Tire experts also suggest rotating tires and wheel balancing on every car's regular service.

The other important aspect is oil, filter change and chassis lubrication. Car experts always advice car owners to change them after every 4,000-5,000 miles or every 3 -4 months, to ensure longer engine life and optimum performance.

It is also important that lights are in working condition. One must check and replace all fuse bulbs every month, to ensure your journeys at nights are smooth and trouble-free. Besides this, we also advise you to check for any fuel leaks by looking for fuel drops below the car every morning, before you start your car. Washing the car and checking the battery once a week will also help you keep your car in perfect condition

Car maintenance is not as technical as it may sound. The maintenance usually consists of procedures that assure cleanliness, monitoring or mechanical performance, and conditioning car parts into their optimum performance in a scheduled basis. If you are a first time car owner, you can ask your sales representative about the maintenance cycle of their product lines so you can also put it into consideration upon its purchase. A car is a liability because of the initial investment it requires but it can become an asset when it turns out to last longer than the manufacturers indicated. Expensive vintage cars are examples of proper maintenance and keeping that enabled their owners to enjoy a more valuable price tag on them compared to when they bought it.

Many local auto shops have lay-away plans that can help those car owners who cannot afford money pay out every time their vehicle goes into the shop. While some car distributors have bundled car maintenance plan as a promotional campaign to attract car buyers. Some auto shops also happen to have special promos and payment plans that can save you from paying too much. There are many deals that you can look out for especially during low car sales months when companies and maintenance shops tend to have special discounts on car maintenance. Car maintenance procedures range from interior cleaning that can include but not limited to interior vacuuming, machine overhaul, filter and tank cleaning, recovery and recharge of aircon freon, cleaning or repair of mechanical parts, oiling and refill, and performance check up.

Exterior cleaning can be pressurized car wash, surface waxing, paint resurfacing, and many others. Accredited auto mechanics and technicians should be the professionals handling the delicate parts of your car so you can be assured of quality and proper car servicing.

To be able to save considerable amount of money in maintaining your vehicle, you should be able to properly gauge the right time in doing its routine check up and maintenance. Having the right schedule means unnecessary trips to the auto shop that could cost you.

Manufacturer recommends a certain span of usage period for your vehicle before it gets the necessary maintenance procedures. Newer car models are more built with energy efficiency and lesser maintenance capabilities. For older models, more trips to the auto shop cannot be avoided due to the lack of lesser maintenance features.If you are contemplating to buy a newer model, it can save you more money in the long run.

Finding a good mechanic in your town can be made easier and faster if you happen to have recommendations or referrals from friends or family. Make sure that the service history of the recommended shop have good track records and has maintained its good reputation for a long time.

Performing spot checks or easy indicators of the performance of your car can help you determine the frequency of car maintenance you need. Checking the oil gauges, the smoke exhaust, and the general sounds made by your vehicle machine can be smart indicators of your vehicle performance.

These are just some of the ways to keep a tab of your vehicle maintenance and tune-ups. You do need to be very careful in selecting the right maintenance schedule and provider so your vehicle investment can fully return its value and possibly last longer than originally designed.

However, in order to enjoy the attributes of any car, one has to ensure that they maintain it properly over time so that it does not develop any problems. One thing that you have to avoid is to wait for problems to become apparent before you can fix them. It is often better to do regular service even when the car is seemingly fine, so that the car is maintained in good condition for as long as possible.

CAR MAINTENANCE -*Shakes, Rattles, Squeaks and Thunks!*

Knowledge is everything, and ignorance is not always bliss.That is pretty much good advice, right? When it comes to driving old reliable why do you see some cars passing you on the highway belching blue smoke, or chugging past you like a steam locomotive on a mountain pass? Are these drivers oblivious to the fact that something is wrong with their vehicle or are they thinking that the problem may just disappear?

Frankly speaking, maintaining our car's health may not be our favourite pastime, but if driving is a daily part of our lives, then it is imperative and our obligation to ensure that we keep an ongoing check on those normal wear and tear items. A little sage advice is in order here.

Today let us take a look at some electrical problems you may encounter from time to time. Are you experiencing starting problems, vehicle stalling, your lights going dim or do not go on? For the most part, electrical problems like these occur from neglect.

There are some simple checks you can do starting with inspecting the acid level in your battery. If your battery is of the non-serviceable type, there may be eye or dot on top of the battery, which indicates battery charge. Note, that this dot only indicates the charge on one of the battery's six cells, and should be only relied upon as a reference point only for your battery's condition. Battery experts will tell you that it is possible for any of the other battery cells to be completely discharged. For the most part, if the dot is green in colour, then the battery should be fully charged. Yellow indicates a partially discharged battery and black means the battery is discharged and needs to be recharged.

Check the sticker on top of your battery to determine the battery's age. The year and month the battery was purchased will be punched out. Today's batteries can last from three to seven years, depending on the vehicle they are installed in, the situations and conditions they are used in, etc. If the battery has exceeded it's normal lifespan, then replace it.

Older serviceable batteries can be topped up have a lifespan of up to ten years, and that is simply because of maintenance. Most important of all is to keep the battery tops dry as water conducts electricity. A damp battery top means that electricity can travel from one terminal to another, resulting in a fully discharged battery.

While you are checking the battery level check to see if there is excessive corrosion on the battery terminals. Corrosion on the terminals will reduce current flow throughout the vehicle's electrical system, especially during startup, when there is a huge current and voltage demand by the starter.

If the battery registers below 12 volts, then repercussions can occur throughout the electrical systems of your car or truck. Starting the vehicle becomes more laboured. The operation of electrical motors, such as those used in door windows, seats, windshield wipers, fans, etc., slow down. Lighting systems go dim and peculiar things can happen when you use your turn signals, such as certain dash lights come on that should not. All these systems start from a 12 volt source and your car or truck's battery is the heart of the whole electrical system.

Which battery lasts longer? As a rule, the lifetime non-serviceable battery should outlast the serviceable type because it is sealed. However, Murphy's Law applies here and sometimes this is just not the case. Good advice is to buy the best battery you can afford.

CAR MAINTENANCE -*Alignments, and Tires, and Brakes!*

You are driving down the road, happily listening to your favorite song on the radio, when it is suddenly overcome by a loud and constant rumbling sound. You immediately realize the problem and, after a few choice words perhaps, pull the car over and confirm your suspicions: a flat tire. Most drivers have experienced the dreaded and unsuspected misfortunes of a flat tire. Many times, you can be successful in preventing this unfortunate event through the proper maintenance of their vehicle's tires.

One of the easiest things a driver can do to decrease their chances of a flat tire and other incidents (such as accidents) is to frequently check their tires. By inspecting your tires at least once a month and ensuring that they are properly inflated and free of stones, glass, metal, and any other foreign objects, you help prevent air loss and accumulating moisture that degrades tire structure which inevitably leads to tire failure.

Checking your tires also allows you to determine certain aspects about the cars performance depending on wear patterns. These wear patterns can be indicators of more severe problems that generally go unnoticed by the average driver. Incorrect tire pressures, tire and steering misalignment, improper balancing, heat excessive buildup, and even suspension neglect can result in issues with your vehicle. By ignoring these warning signs, tires can become bald and excessively worn, causing sluggish driving, skidding, and even the unnecessary accident.

For example, if your tires are not balanced properly, you may experience constant vibrations as you drive, which can lead to wheel mis-alignments, bent wheels, and inadequately operating brakes in addition to the rapid and uneven wear of tires and loss of air pressure. In addition, by checking the suspension you are able to identify early problems with joints, bad brakes, shock absorbers, springs, which can drastically improve weight transfer, spring rate and damping as well as correcting other issues so new parts are not needed as frequently.

In short, by having your tires, brakes, shock absorbers, drive train, steering, and suspension systems regular inspected by a qualified professional, problems can be diagnosed and corrected so you can have a safe and pleasant driving experience.

There are things that you too can do to prevent problems with your car such as checking cold tire pressure with a gauge once a month, inspecting the tires, and following the owner's manual and a tire safety handbook. However, if you do start experiencing problems, it is important to consult a professional or get your vehicle to a service center as quickly as possible to prevent additional damage.

Many people feel that the cost of spending more money on better tires or paying for professional services is far more than is necessary for the upkeep of their vehicle, but is that really the case? For a wheel alignment you will generally spend less amount (and many times come with special offers or additional warranties).

While this may seem like a lot, the results include improved tire wear for longer lasting tires, reduced strain on suspension, and better gas mileage, all of which will increase the amount you spend for maintenance on your car over time.

The car is often seen as the lifeblood of the working person, and to many an extension of themselves, so it is natural for a driver to want to guarantee that they keep their vehicle a "well-oiled machine" for as long as possible. Regardless of the trouble your vehicle is having whether it is with the alignment, balance, tires, brakes, suspension or even something as simple as an oil change, trained professionals are ready, able, and willing to help you with whatever problem you may have.

What to Do If Your Car Brakes Are Squeaking?

Performing a quick and simple test on your car breaks may save you lots of money and headaches. Bring your car to a speed of about 30 MPH, then step on your brake pedal very slightly and look for a brake pedal pulsation or any kind of unusual brake noise.

A pulsation on the brake pedal indicates that your rotors are out of round. Out of round rotors are not uncommon and usually the cause of this is excessive heat and cold. The heat produced by the friction created between the brake pads and the rotor is huge, hotter than an empty cooking pan with a full flame. Therefore you should avoid driving into a puddle of cold water after long drives.

Brake noise or brake squeaking suggests a potential brake failure that needs to be addressed promptly. It could be a critical safety issue and you could be putting at risk not only your live but also does of your loved ones.

You see, now-a-days most brake pads manufacturers are installing "squeaking" metal device on their pads with the sole purpose of warning the driver that the brake lining is wearing thin so that the brake pads can be replaced before they damage the brake rotors.

Disregarding this annoying squeaking noise will transform into a grinding noise which means that the lining of the brake pad is completely gone and the brake pads' back plate is now grinding against the brake rotor. At this point, you may still be able to stop the car; although, the brake pads' back plate is destroying your rotors with each turn of the wheel and every time you apply the brakes.

Ignore the noise a little longer and the pads' back plate will grind through the rotor causing you to lose your brakes instantly and perhaps causing not only an accident but also the loss of human life. Do not take unnecessary risk and have your vehicle checked out by a professional mechanic as soon as you hear any kind of brake noise or at any time you suspect a problem with your brake system.

Of course, this irritating noise may be just a rock stuck between the brake shield and the rotor, which could be fixed simply by extracting the rock. The brake squalling could also be caused by an inferior quality of brake pads that may have glassed the rotors. If this is the case, it will be necessary to resurface the rotors and install a new and good quality brake set.

Whatever you do please do not experiment with your brakes by adding any kind of oils or lubricants on the brake pad linings to stop the squeaking as others have suggested. Adding any type of lubricants to the pads could cause slippery brakes and possibly an accident. Also, oils and lubricants can damage, dissolve, harden, soften or react on brake pad lining making it unsafe.

Remember that common sense is your best tools to prevent accidents. Trust your common sense and you will know when to take appropriate action. Keep in mind that the brake system is your most important device on your vehicle. Do not try to save a couple of dollars on a cheap brake set that could cost you dearly in the future.

CAR MAINTENANCE -*Checking the Wheel Bearings!*

Firstly, what is a wheel bearing? If you have ever seen that metal ring that encloses what looks like steel balls, you know you are looking at a bearing. The purpose of a bearing is to allow as little friction as possible when the wheel is spinning during motion.

To check for a bad wheel bearing without having to drive, place the car in neutral and check that the hand brake has been disengaged. The car has to be lifted to check for free motion of the wheels whilst you do this test. Spin the wheels a couple of times to check their movement and motion. They should both be spinning freely without too much input from you.This is an indication that the wheel bearing is fine. If one of the wheels needs more force and pressure to spin, it means that the bearing is worn or faulty.

You know that you need to check or possibly replace the bearing if it starts making noise whilst it drives.

The only problem then will be that you will have to troubleshoot what is causing the noise, since the car has to be in motion to produce that sound.

The noise that you should be listening out for is a dull and blunt sound. This noise sounds like a bunch or marbles rolling on brick paving. Remember that the wheel bearing is, in a way, a bunch of steel marbles in a metal ring.

Checking which side the faulty or worn bearing is requires for you to get in your car and take a quick drive down the road. Sway the car from left to right as gently as you can to listen out for the noise. If you turn or sway left and no noise is heard then the bad bearing could be on the right side. If the sound gets stronger and louder as you drive faster, you know for a fact that it is the wheel bearing or ball bearings that need to be replaced with brand new ones.

Your car is known for making all sorts of sounds. The chances are high that the dull sound you are hearing is from the tyres going over bigger pieces of rubble. Listen carefully for all the sounds, especially the unusual ones.

Before making any assumptions and deciding to take the time to replace the wheel bearing, take the car in to a professional. A wheel supplier or a car specialist will know the exact sound to be listening for.

CAR MAINTENANCE -*How To Change Your Car's Oil!*

There are many places you can go to get your oil changed. Most of the places are not too expensive, but they are often busy, and it is hard to get to the shop when you are busy working. If you are interested in changing your car's oil yourself, there are things you should know. First of all most cars are pretty much the same in this regard. If you have changed one car's oil you can probably change it in most cars. While the oil pan may be located in a different area of each individual model it is generally on the bottom of the vehicle. Below are a few tips that will help you make this task easier.

You will need at least six quarts of oil. Preferably you should use the same type of oil as you had in the engine before.

You will also need to purchase a new oil filter designed specifically for the model of your car.

During an oil change most people also change the air filters on their cars.The air filter tends to become dirty in about the same time as the oil needs changing, especially on older model vehicles.Therefore it may not be a bad idea to purchase an air filter as well.

If you have not done any prior work on the vehicle at home, you will need to purchase a car jack or tire racks. These racks allow you to drive up on them to elevate the front end of the car where you will be working. If you are using a jack or the racks you will need to place blocks behind the other two tires to keep the car from rolling because the parking brakes do not always hold the grip reliably.Make sure that the parking brakes are on and that the transmission is in park position. You need to elevate the car so that the oil will run out of the oil pan completely as well as make it easier for you to reach the oil pan.

Prepare a canister or a pan to catch the oil in. Most auto parts shops including those online will have oil pans that you can drain the oil into. These pans have a special design to make disposing of the oil into an oil drum easy.

You will also need an oil filter wrench.This wrench must be the correct wrench for the size of filter you have. There are many sizes depending on the type of car you have.Another tool you will need is the wrench for the oil pan nut.This nut has to be removed to let the oil escape.This is usually a metric or standard wrench depending on who manufactured your car.The best tool for this job is a socket wrench.

Once you have all of the tools and other items you will need to locate the oil pan. It is best to look in your car's manual for this information. Keep plenty of towels on hand as well.

First, loosen the nut on the oil pan, making sure that you are prepared for the first gush of oil by having the waste pan underneath the oil pan.

Next, locate the oil fill cap and remove it. Some manuals may tell you to do this first. It has been found that this creates more of a gush if you wait to remove the hexagonal nut after, so the potential for a spill is higher.

Once you have gotten the oil flowing out of the vehicle you can then set a bucket just under the oil filter. There will be a little oil drainage from the filter so having the bucket under there will help keep you from causing a spill. Once you have removed the old filter you can then screw on the new filter following the instructions on the box.

Now that you have removed all of the oil from your vehicle you will need to make sure that you replace the hexagonal nut on the oil pan and make sure that the oil filter is screwed on tight.Then you can begin replacing the oil. It is best to fill the engine with at least four quarts of oil and then let it settle. Check the engine oil level and then add a quart until it is full. Check with your manual on the exact amount of oil required for the engine.

CAR MAINTENANCE- *Car Servicing!*

Men treat their cars as precious as their girl friends.They do a lot of upgrades and accessories just to make it look good.They do everything to enhance the beauty of their cars. Mostly they upgrade the engine into car racing type. Add accessories like bumper, skirts, tented windows, turbo engines, mags, wheels, and a lot more.

They spend money for its body beauty but also you need to consider the maintenance of its engine.

Car Servicing is one of the effective ways of maintaining the quality of your car. Autos need to undergo into regular check up to avoid any repairs. With proper car maintenance it will last for several years and you will be able to use it when needed. When you buy a brand new car,most car companies have car servicing offers included to the car warranty. You need to follow the car servicing maintenance being schedule according to its miles. Auto owners must follow the advice of the car companies because they know better. If it is your first time to own a car, you must ask your friends or people whose expert in car maintenance like the car mechanic.

You should not take for granted the maintenance of your car.Human being needs regular check up to their doctors while auto needs regular tune up and check up from auto servicing.

Never do the car maintenance by your own unless it is a minor maintenance and you had experience with cars. But if you have no experience at all, do not risk any of it and just go to the nearest car servicing in your area. Listed below are some maintenance tips you should consider in taking care of your cars:

• *Regular Car Service* - It is important you follow the car service schedule according to exact distance or miles or time period. Most of the time, car companies scheduled it for you. It usually happens in first 1-3 years of owning the car.

• ***Proper Oil Change of Engines*** - Oil change should be done once a year or depending on it running miles. Follow the mechanic advice. With proper engine oil change guarantees a long engine life.

• ***Check Fluid Level*** - It is something can be done with the car owner but if you do not know how to do it ask for your mechanic helps although it is very easy to do it yourself. The essential fluids like coolant fluid, Brake fluids, battery fluid, and transmission fluid should be checked and maintained on a regular basis.

• ***Check Brake and Tyres*** - These are the commonly used factor of your car so it advisable to check them every day if they works fine before using the car. Check if the break pressure is on the right level. Check if tyres are performing well with the right amount of air.

• *Car Body Maintenance* - You should also maintain the looks of your car. See if it needs body repaint. Check the minor parts such as side mirrors, headlights, bumper, doors, seats, and other body parts of the car. It is not necessary to have brand new body parts every day but just make sure it still working fine for you.

Importance of Car Servicing

With people's lives becoming busier and busier the smaller things tend to get pushed to the back of the mind. This is especially true when it comes to maintaining your vehicle. Many of us use our cars or bikes on a daily basis without giving them a second thought. The only time we do tend to take notice of them, is when they have decided not to work on that crucial day you need it to.

This highlights the importance of not only regular car maintenance, but regular professional car servicing. Like every other technical item we use on a daily basis, our cars need some care and attention to ensure they keep going for as long as possible. If you use your car to commute to work or take your children to school, you should make your car servicing a high priority.

Many people believe a car service is simply unnecessary and will avoid it at all costs as they feel it can be a waste of money. However, the truth is that a regular car service could potentially save you thousands in repair costs in the future. By having you car regularly looked at by a professional and tested using specialist equipment, you may avoid major cost implications by solving the minor problems before they turn to major issues.

When organising to have your car serviced, it is vital you choose a reputable company who is familiar with the servicing the make of car you have. If you find a garage which has trustworthy awards such as the VOSA survey 'Green Light' award, you can be assured they are adequately trained to carry out such a professional service. When putting your car in for service, you should express any concerns you have or anything unusual you have seen or heard. This will also highlight any minor problems before they turn into a major problem.

You should also aim to find a local garage, for example if you are in Paris, opt for a garage in Paris. It would not make sense to choose a garage which will take you an hour to drive to as it is likely you will need them when your car is not working.

Car services will also tell you an appropriate time to trade in or replace your older vehicle. If you are finding every time you visit the garage your car requires more work, it may be more cost effective to trade to a newer model with fewer problems. Car services will also help you when it comes to selling your vehicle. It is a well known fact that you should steer away from buying cars with no service history. If you have a full service history at regular intervals, it will make selling your car a much easier job as there will be high demand for it.

CAR MAINTENANCE -*Different Fluids Cars Use!*

Keeping the different fluids in your car clean and full, is part of your basic car maintenance that you do not want to neglect. By keeping your car well-maintained you can help prevent breakdowns or future problems. It is not uncommon to see an engine go well over 200,000 miles if it has been properly maintained. If you do not know where to find the fluids in your car you can always refer to your owner's manual to find them, and see what maintenance is required.

Here are the different fluids your car uses. It is also important to know what these fluids look like so you can identify them if you have a leak.

Engine Oil (Dark amber or brownish color) - The motor oil in your car keeps all the engine parts working and lubricated, without this oil your engine would seize up and stop running. Check your engine oil at least once a month and make sure it is filled to the mark on the oil dipstick. Make sure you put the proper type oil in your car by checking with your owner's manual.

Power Steering Fluid (Goes in clear but turns darker with usage and age) - This keeps your car turning easily and effortlessly.

Automatic Transmission Fluid (Usually a reddish or pinkish color) - This important reddish, fluid allows your car to move backwards and forwards easily and smoothly by shifting gears while you drive. A manual transmission may use several different fluids. Check your owner's manual to make sure you are using the right fluid.

Brake Fluid (Goes in clear but turns amber with usage) - There is a reservoir for the master cylinder up on the firewall of the engine compartment where you can check the level of fluid. Keep it up to the full mark. The brake fluid allows you to stop your car. Without brake fluid you literally have no brakes.

Engine Coolant or Anti-freeze (Florescent-colored green or orange liquid) - Keeps the cylinder heads and engine block cool so your engine does not overheat and seize up.

Rear End Oil (Dark amber or brownish color) - This allows your drive axle to turn freely and smoothly.

Windshield Washer Fluid (Usually a blue color) - Used to help keep your windshield clean and streak free for good visibility.

It is important to maintain the proper levels of fluids in your car for safe operation and in preventing trouble down the road. If you notice any signs of these oils or fluids on your garage floor they need your immediate attention before causing serious problems and damage.

CAR MAINTENANCE - *Car Tuning!*

All of us dream of getting good at something we have been passionate about since we were a kid. One such thing would probably have to be car tuning. There are a number of enthusiasts out there that are involved with doing this on their own vehicles. After all, considering how simple and straightforward it is, it would be unwise to have it done for you by someone else. You might want to consider the option of learning more about the concept before you can actually try it out on your own. Listed below are some ways in which you could do this.

Researching about the car

One way that you can significantly improve your knowledge about car tuning would probably be if you research about your vehicle and perhaps learn more about it in the process. This is an important thing to keep in mind about, since you would not want to end up trying something out with half knowledge. Some people spend as much as a year or more learning about their car before trying something on their own. You too could do the same to get better at it and perhaps learn to understand the simple stuff that powers your vehicle.

Getting involved with the maintenance

Car manufacturers provide a maintenance schedule for the reason that car owners know how to take care of their vehicle. Hence, in this manner, car owners can get familiar with car tuning and take care of their car at the same time. This is something worth looking into, even if you are not particularly interested in tuning your car.There are multiple benefits of taking care of your vehicle like intended. It is quite clear that you gain quite a bit by putting in some effort and taking care of your vehicle.

Going to online forums

A shortcut to getting knowledgeable about car tuning would probably be by going to some good online forums and learning more about the vehicles from some of the people in these forums. This too has yielded some good results in the past. You would definitely stand to gain from this and trying it out is highly recommended. Many of these forms are interesting places to go to in order to learn more about your car and also what a particular component does for your machine. It is definitely worth looking into it for any kind of car lover.

Although the online forums are a great idea, do be warned that it need not necessarily be the best one out there. You might want to be a little cautious about this and consider the website that you are going to in order to learn more about car tuning. It is important that the information you obtain is trustworthy and dependable. After all, you would not want to simply complicate things by taking into account something that is not necessarily good for your car. Hence, spend some time and figure out all aspects before you can actually go ahead and try something out.

Essence of Car Tuning

Our car is the most precious possession of ours. For real you take utmost care for it. But what will you do with the pot holes, speed bumps and rough roads. They are part of your every day driving. Your cars comes in contact with all these every day. That is the reason why you need to keep it well tuned. In fact if you also got a brand new one, then also you need tuning. It will help your car to serve you for a longer period of time. Basically this is a common fact that something which has moving parts tends to break very easily. That is also another reason to send it for tuning at regular intervals.

Car tuning includes several aspects like engine tuning, drive trains tuning and many other parts. Thus modifying these parts helps a car to give higher performance. Tuning is important for every car. But it is very much needed for the modified one. Sometime some of the modified cars are damaged due to continuous use. It is also important to see the modified part is not harmful for them.

Your car will give higher performance if it is properly maintained. The main aim of tuning is to improve the handling and performance of it. Basically the manufacturers of car develop car in bulk. So they concentrate on type and style. In such cases the quality of the car is sacrificed. So by tuning your car you can adjust different parts of the car according to your driving capabilities.

Car tuning helps in increasing the speed and power of your car. In order to increase the power an efficiency of your car you need to install turbo chargers and new cooling systems in the engine.

More over the price of fuel is soaring high. Here also car tuning can help you out in saving your money. It is like a medication on them. It helps in making your car more efficient when it comes to mileage. Thus your car will consume less fuel and will save a lot of money.

Thus proper maintenance of your car will help it to serve you for a longer period of time. More over it is your dream car, so you also want your car to remain as fresh as a new one. So car tuning is the best solution for that. However if you are not experienced or are not very mechanic it is better to take the help of some experts or professionals for your car tuning. This is to ensure that every thing is done properly. How ever the main thing is that you need to maintain your car, other wise your all efforts of car tuning will go in vain. It is not very difficult task to find a provider of such services.

CAR MAINTENANCE- *Some Main Tools!*

There are some basic car maintenance tasks you can do by yourself. You only need to equip yourself with some main tools. With these tools, you do not need to call a road service for a simple maintenance task such as fixing tire or replacing the fluids. It will be able to help you save money in a long run.

Jack, Wrench and Tire Iron

You have to always remember to keep the car jack, wrench, and tire iron while on the go. It will be very beneficial if you get flat tire. Place the jack under the car close to the flat tire for lifting the car up so that you can easily remove the tire. Wrench will help you remove wheel lugs or bolt while the tire iron can help you secure the wheel lugs back to the spare tire while on the wheel. In addition, wrench is useful to remove oil tank bolt. You will use it in case you perform oil change.

Funnel for Fluids

Funnel will help you perform the maintenance task such as replacing fluids like transmission fluid, oil, and brake fluid. This is one of the most important parts of car maintenance. With this tool, you will be able to avoid the spillage while filling up the fluid reserves.

Tire Pressure Gauge

Tire pressure gauge will help you keep air in the tire at the appropriate level. Driving the car with low pressure air will reduce the efficiency of the fuel since under-inflated tire can make the engine work harder. Later, it can affect the steering. Therefore, before driving, you have to make sure that the tire pressure is at the appropriate level. To measure the tire pressure, you only need to place it over the nozzle of your tire.This tool shows you the reading on the tire pressure through the small poles which looks like a thermometer.

Battery Charger and Air Machine

Car battery charger is beneficial in case you are running out of the power since you accidentally leave the lights on. To solve your problem, you only need to attach the handles of charger to the according red and black connections and then switch the device on to charge it. Instead, you can also get the jumper cables and connect the handles to another car. In addition, you can also have an air machine for filling up the flattened tire.

CAR MAINTENANCE- *Radiator and The Engine!*

If you believe in the concept that all good things in the world come in pairs, then you must be wondering who could be the companion for your car engine. Stop beating about the bush, it is the *radiator*. A radiator is an integral part of the engine setup that runs your vehicle. In fact, these two parts of your car are mechanically linked and even though the engine might be the fiery, hot-headed type, it isthe radiator that has a soothing influence over it and that is what defines their relationship as well.

If you are not familiar with the major parts in your car, let me introduce you to what is called the Car Radiator all over the world. When you pop-open the hood of your car, the first part that immediately follows the front grille is the radiator. In fact, a radiator itself resembles a grille or a mesh of lean pipes forming a square shaped part that has an opening at the top and a fan located just behind.

The location of the radiator is of extreme importance allowing it to perform its main function i.e. cooling the engine. Let us understand how it manages to do that.

As you would know, all internal combustion engines produce power by burning fuel and this power drives the wheels of your car. A mixture of air and fuel is burnt within the cylinders of the engine and the resulting explosion pushes the pistons of these cylinders that are connected to the crankshaft. Constant periodic explosions inside the cylinders of the engine set the crankshaft in motion. The crankshaft then converts the linear motion of the pistons into rotational motion that travels through the drivetrain of the vehicle and turns the wheels.

But you do not need to go into all that technical stuff. The fact is that burning of fuel heats up the engine and the same effect is created as a result of friction between the moving parts of the engine. Although engines are made to withstand high temperatures and pressure, if this happens on a consistent basis then engine damage is imminent and that means either an end to your vehicle or a big hole in your wallet.

The truth is that most car engines cease at high temperatures and a consistently heated engine will definitely have damaged internal parts. As you would know about badly maintained engines or malfunctioning parts, they lead to loss of efficiency and power and less MPG means another hole in the wallet and this one will not mend.

Automotive engineering theories suggest that the engine performs at its best when maintained at standard conditions of temperature and pressure and that is what brought about it's pairing with radiator.

A radiator essentially contains anti-freeze or coolant that is circulated through the engine to keep it cool. The anti-freeze absorbs the heat produced by a running engine while the cold antifreeze cools the engine. The coolant that gets heated in the process is circulated back to the radiator where it flows through the narrow pipes of the radiator grille. As the vehicle moves, the air entering through the grille of the car cools down the heated anti-freeze contained in the radiator.The fact that radiator pipes are narrow means it takes less time for the anti-freeze to cool down. If the outside air is not enough, a fan located behind the radiator does the job automatically thus keeping the temperature of the engine under control and preventing it from overheating.

The engine delivers power to run the vehicle but it is the radiator that maintains the health of the engine. Since the effectiveness of the engine also depends on the conditions it operates in, a well performing radiator with good levels of anti-freeze in it can make the difference between an engine and an effective engine.

If you want your vehicle to perform at its best, you have to take great care of its engine and an effective radiator can contribute immensely in this regard.

Maintain recommended levels of coolant in the radiator and service the engine on periodic basis to bless this pair and in turn you will get a blessed vehicle.

CAR MAINTENANCE -*Common Signs of Car Radiator Failure!*

The radiator is an important component in your vehicle and a primary part of your car's cooling system. Its job is to keep the engine from overheating, which is vital for optimal safety and performance. If your car's radiator begins to show signs of needed repair, it is wise to have it serviced by a professional and licensed mechanic as soon as possible. Catching a small problem early on is a cost-effective method of routine car maintenance and repair. Continue reading to learn the most common signs of car radiator failure, and test your vehicle to assess its current condition.

Low Coolant

If you begin to notice that your coolant is always running low, or your "low coolant" light comes on, it may be due to a radiator leak. Although it can be tempting to simply refill and forget about it, it is important to put your schedule aside and have your radiator inspected by a professional mechanic. A radiator leak can be dangerous for many reasons, so it is necessary to repair them if one exists.

Coolant Leak

If you notice coolant on the ground beneath your vehicle, you have a coolant leak. Coolant leaks are caused by a leak in the radiator. Radiator fluid, known as coolant, flows through the engine and the radiator, so if it is dripping onto the ground, it is a sure sign that the radiator has a crack or opening somewhere. A professional auto repair shop can accurately locate radiator leaks with a specialized test using pressure and dyes.

Discolored Coolant

Radiator fluid should be yellow, red, or green at all times. However, when the cooling system begins to go bad, the fluid can start to turn to rusty or oily colors, such as black or brown. Overtime, this oily fluid turns into sludge inside the radiator, which prevents the coolant from flowing properly. When this happens, it slows performance and reduces efficiency. Unfortunately, a radiator will need to be replaced if sludge gets inside.

Overheating Engine

Since the radiator's job is to regulate the temperatures produced inside an engine, it is no surprise that an overheating engine is a very common sign of radiator failure. If the engine overheats just one time, it could be due to something minor like being low on coolant. But if your engine is overheating regularly, it could be a more serious repair. The more times you allow your engine to overheat, the more damage is done to your vehicle. It is important to get them repaired as soon as possible before they can get worse.

CAR MAINTENANCE -Engine Overheating?

If your car engine starts overheating, it could result in serious and expensive damage to the engine as mentioned earlier and your car needs to be serviced immediately.There are several reasons that your car may be overheating.

If your car is low in coolant, it may be overheating. A car engine's cooling system depends on coolant to circulate and remove heat from the engine. If there is not enough coolant to circulate, heat will build up and the engine will overheat. You should check your coolant level. You should always check the coolant with the car is cold. This can be done by looking to the right of the radiator for the overflow tank. It is white plastic but you should be able to see inside to decide if there is enough coolant. There are markings on the side that indicate the low and high levels. The engine takes a 50/50 mixture of coolant and water. You can now purchase a premixed coolant that is ready to pour, from an auto parts store. Unscrew or pop off the cap of the plastic overflow reservoir and add the mixture until it reaches the full mark. Once you put the cap back on tightly, if that was the reason your car was overheating, the problem should be solved. This is probably the most inexpensive and easiest way to to check the reasons for engine overheating.

If you refill the coolant and your car is still overheating, it may be an electric cooling fan failure. If the fan fails to come on, it can cause the engine to overheat. The cooling fan draws cooler air through your radiator when your car is not going fast enough to bring it through the front. To test this, let your car idle long enough for the engine to heat up. Keep an eye on the temperature gauge. When it starts going into the red, look under the hood to see if your electric fan is running. If it is not, there may be two reasons - a bad electric fan or bad radiator fan switch. Sometimes the fan motor will burn out and the fan will not come on at all. You can test this by finding your radiator fan switch and disconnect the wiring. Get a jumper cable and insert it into both contacts - your fan should come on. Another way to test the fan is to turn on the air conditioning. Most cars activate the cooling fan at medium or high speeds when the air conditioning is turned on. The radiator switch is a switch that tells the cooling fan to come on when the coolant reaches a certain temperature. To test this, Disconnect the wiring harness and then run a jumper wire or cable across the harness contacts. If the fan comes on, you will need to replace the switch.

If your car is overheating, it could be that the thermostat is not opening. The most common symptom of a failed thermostat is overheating at highway speeds. The engine may be able to stay cool at low speeds because it is not working as hard and not creating as much heat. When you drive at highway speeds, the engine needs lots of coolant running through it. If the thermostat does not open, there is not enough flow to keep things cool.

A broke fan belt could be another reason for the engine overheating. A fan belt operates the engine's cooling fan. Electric cars may not have a fan belt. If the fan belt is broken, it can be easily replaced.

A clogged radiator may be the reason your engine is overheating. If your car has more than 50,000 miles, your radiator may be getting clogged. You should flush your radiator every year.

A hot engine can do damage to itself, so even if the engine is not overheating fully, it can still be causing damage. You should check your oil regularly to make sure there is adequate lubrication to the engine so that the engine is not running against dry friction.

Today, more than ever, cars are reliant on a car's electrical system to run properly. It used to be that voltage swings in cars, or even a thrown alternator belt, were things that could be shrugged off temporarily, and you could still get to a service station or your destination with only mild discomfort. Now, an under or overvoltage condition can cause your car to go into limp mode or even shut down completely. The sensors and electronic modules in today's cars are sensitive to a specific voltage range, and the wrong voltage can send a wrong signal to the ECU, causing problems. And with the tendency of many owners to add aftermarket accessories to their car, it is more important than ever to look after this system.

Basic care for your car's electrical system is actually easier than caring for the mechanical components in the car. This is because all you need to do is visually check for corrosion or unusual signs in the system's components, of which there are but the battery and the alternator. The starter is also one component of the electrical system but it is rare for it to be cause of an electrical fault.

The most obvious component to the electrical system is the battery. Often, it sits on a tray in the engine compartment. In sports cars or cars with sporting purposes, the battery can be located at the rear of the car, where it helps with weight distribution and is isolated from temperature extremes. Wherever their location, batteries should be securely held down as a loose battery is a dangerous component to have flying around inside the engine. Poorly secured batteries will also be subject to more vibration than necessary, which could damage the plates inside, thereby shortening its life, often drastically. Check the terminals every couple of months for corrosion or loosening. If you see a whitish residue on them, remove the terminals and clean them with soap and water. It helps to put an anti-corrosive coating on them to prevent the condition from returning. Plain grease works. Minimizing heavy loading will also help your battery's life. Many owners start the engine with

A/C switched on. Or a high-powered sound system or their HID lights for that matter. With these turned off, the load on the battery, as well as the starter, will be less when you start the car.

The alternator is driven by a belt connected to the engine. It generates electricity when the engine is running and tops up the battery to the proper voltage. Normally, you do not need to do anything to an alternator. As with the battery, visually inspect the connections for signs of corrosion and clean them accordingly. Important note: Always disconnect the negative terminal of the battery when cleaning the electrical system. You may damage an expensive component if you short a circuit inadvertently. Unless you have made some changes to the engine compartment in terms of ducting or intake tract mods, you will not need to check for obstructions to the alternator's airflow, the lack of which could cause overheating.

Lastly, check your belts for proper tension. A loose belt may cause slippage which can cause an undervoltage condition from the alternator. An overtightened belt on the other hand will stress your bearings shorten their life. Check the underside of the belts for cracks. If you see any, replace them as soon as possible.

There you have it. Simple checks to your car's electrical system that will help it give long and reliable service.

CAR MAINTENANCE -Exploring Your Car's Fuel System!

When consumers visit the gas station, the most common thought on their minds is the price they are paying per gallon. Few people think about the process by which gasoline travels from the tank into the engine. This will be discussed here.

If you are familiar with the workings of your vehicle's engine, you already know that fuel and air mix in each cylinder's combustion chamber. Within, the mixture is compressed as the valves close and a piston rises. A spark plug sits on top of the chamber and generates the spark needed to ignite the compressed mixture. That causes a small, contained explosion within the combustion chamber. The energy and the expansion of vapors pushes the piston downward, which aids in the propulsion of your car.

But, how does the gasoline make it into the combustion chamber in the first place?

The Gas That Flows Through The System

Despite what a lot of consumers think, no two fuels are exactly the same. To be sure, all of them have the same compounds. Moreover, they have similar additives and detergents (for preventing olefin deposits on your fuel injectors). But, each poses a slightly different level of volatility, which can loosely be defined as the ease of vaporization.

If fuel vaporizes (i.e. burns) easily, your engine can operate more efficiently. However, if it burns too easily, the mixture within the combustion chamber will be too lean. Similarly, if the gas does not burn easily enough, the mixture might be too rich. Given the high operating temperature of your engine, the level of volatility of your gas is important.

The Storage Facility

Your gas tank is where the fuel remains until it is needed by your engine. In most vehicles today, the tank is located toward the rear. Part of the reason is due to space limitations in the front. Most modern tanks are equipped with baffles that prevent splashing. If you are able to hear splashing, that usually means that they are broken.

The Network Of Hoses

Leading from your gas tank to your engine is a small network of hoses and lines. The former are made of rubber; the latter are made of steel. Eventually, they will need to be replaced as a result of normal wear and tear. The rubber hoses must be replaced with the right type of hose to avoid deterioration. The steel lines must be replaced with steel.

Pumping The Gasoline

If you are driving a vehicle with a fuel injection system, it has a electric fuel pump. Cars with carburetors usually have mechanical pumps. Though it was not always the case, the fuel pump is normally built inside the gas tank. When you turn the key in the ignition, the pump receives an electrical signal.This signal generates the necessary pressure to push gasoline out of the tank and through the fuel lines.

The Fuel System's Crossing Guard

The fuel filter is arguably one of the most important components in the system. It helps to prevent dirt and debris from accessing - and clogging - your injectors. If the filter develops a clog, the pump is forced to work harder. That eventually causes it to burn itself out.

What may surprise many drivers is that their vehicles actually have two separate filters: one in the gas tank and the other in the line that leads to the injectors. It is the latter filter that needs to be periodically replaced.

Your vehicle's fuel system works seamlessly to keep your car on the road and operating efficiently. But, parts occasionally fail. Whether it is the tank, hoses, pump, or filter, make sure you are using high-quality OEM-certified replacements.

Improving Vehicle Performance With Fuel System Cleaning

All modern vehicles come with fuel injection systems, so it is a topic we all need to know something about. All new cars and trucks sold over the last 3 decades or so have come with fuel injection systems.

The fuel injector is a valve that delivers the gas or diesel to the right place, in the right amount, at the right time; to be mixed with air and burned in the engine.

So how many fuel injectors does your car have? There is one for each cylinder. So four, six or eight for most folks. Some vehicles have 10 or 12 cylinders. The engine control computer makes adjustments to the injector as it monitors the engine and other sensors. Fuel injectors are a pretty sophisticated part.

Fuel injectors like any other part of the car do need to be maintained and cleaned. What is the benefit? In order to work right, the injectors have to deliver the fuel at a precise pressure at a very precise time. It needs to be sprayed in a particular pattern as determined by the engine design.

Over time, varnish can start to build up in the injectors, effecting the pressure, pattern and timing of the fuel charge. The result is that the gas or diesel does not get burned as efficiently as it could.That robs performance and wastes expensive gas or diesel.

What about dirty fuel - how does that affect the fuel injectors? The fuel injectors are the last stop in the fuel system. It starts at the tank.Truly,the best way to keep your injectors working well is to use high quality fuel. Its real tempting to shop for bargains with fuel prices as high as they are, but major brands have better detergents and additives and deliver consistent quality.

What about a good fuel filter? The filter is the next device in the fuel system. Its job is to filter out the dirt and rust that collect in the fuel tank. If it is clogged up, the dirt will by-pass the filter and head upstream to the injectors.

It is important to replace the fuel filter when your manufacturer recommends it. That is part of a comprehensive fuel system cleaning.

There are different kinds of fuel injection systems. Port fuel injection systems, the kind most gasoline engines have, operate at 60 pounds per square inch. The injectors for the new gas direct injection engines we are starting to see require 10 to 30 times as much pressure. And some diesel engines for passenger vehicles have injectors that operate at 30,000 pounds or more per square inch. There is no room for dirt and gum in a precision part like that.

There are a lot of good products available that can clean fuel injectors. They are best used to prevent fouling in the injectors. Many can not clean a seriously gummed up injector – that requires a professional deep cleaning. But putting the cleaner in the fuel tank after you have had a mechanic take a look at your fuel system will help keep it clean. Be sure to read the label for directions.

Ignition system is responsible for making ignition occur within the combustion chamber of each cylinder head. Without ignition,we would not have combustion and therefore could not run an engine.

After your car has been started, the ignition system keeps the spark going to each individual spark plug, timed in conjunction with the time that the piston will reach the top of its compression stroke so that the fuel can burn. It does this by passing current from the alternator to the ignition coil where it is amplified and passed to the distributor. The distributor dictates when the spark should travel down to the plug, and when it does, the current flows through the wire and jumps the gap between the spark plugs electrodes creating the ignition needed for the engine to run.

Any electrical failure in this system will result in a car that does not run or runs very poorly and erratically. Be careful while working on the ignition system as the coil can supply up to 40,000 volts in most cars while high performance coils can supply nearly 200,000 volts. That is more than enough to give you a good shock and possibly seriously harm you. Remember to work smart by working safely.

CAR MAINTENANCE - *Car Head Lights!*

Car headlights or headlamps are a very important car accessory which enables you to drive safely at night and also during extreme weather conditions. Beams of light help to illuminate the road ahead and provide visibility for the driver for safe and easy navigation of the vehicle. Hence it is important that these lights have to be cleaned and maintained in good condition for your safety and also to increase the life of these lamps.

If you notice the headlights to be cracked or broken, make sure that you replace these immediately. Even if it is just a slight crack, water and dust particles will enter and thereby dim the light emanating from the lamp. The lamps will also give out a sharper glare which is very inconvenient for other drivers on the road. But if you are not able to replace the car headlights immediately, make sure that you seal the crack using a good quality resin as a temporary fix.

Clean the headlights regularly so that the external dirt settled on the surface will not cause dim light. You can either use soap and plain water to clean these but ideally, it is best to get a headlamp cleaner from online stores or your car accessory store which has been designed to keep the glass clean and spotless. Also ensure that you get an annual maintenance done from a mechanic so that any faulty wiring can get corrected and also ensures that the lamps are facing the right direction. It is natural that the bulbs inside can get fused; the ideal thing to do would be to replace both the bulbs even if only one is fused. It is also important that you replace the bulb if you notice the light to get dimmer rather than wait for it to fuse and go out completely.

If you notice any condensation on your car headlights, then this can be removed easily at home itself. All you need to do is to drill a fine hole at the bottom of the light after carefully removing it first. Keep it in a dry place so that all the moisture dries out and then reseal using a top quality silicone sealer. You can place the headlights back and continue using these for a long time to come.

Car headlamps have to be maintained well as these are of utmost necessity for safe driving. Hence keeping these tips in mind will ensure that your car lights glow bright always.

CAR MAINTENANCE – *Batteries!*

What could be more unpleasant than when you rush to work, and your car battery dies. Quite sure almost everyone has at least once had a situation like this. This can happen for two reasons: either because of negligence when we leave the lights on over night, or when we fail to properly maintain the battery and the condition of the battery erodes over time until it completely fails. In the second case, we usually fall for an urban myth - a misconception about lead-acid batteries in cars.

Below are some myths about car batteries;

- Storing the battery on concrete pavement will discharge it
- Driving a car will fully recharge its battery under any circumstance
- On very cold days ignite the headlights to "warm up" the battery before starting the engine
- Lead-acid batteries have memory
- A higher capacity battery will damage the car

- Once formed, the car battery can not change its polarity
- A car battery will not lose power during storage
- Defective car batteries will not affect the loading or starting

Some Facts About Batteries

- Batteries in a wooden case (about 100 years ago) would really discharge when put on concrete. But since they are securely sealed in plastic containers now, there will be no unintended leakage.
- It takes quite some hours of constant driving at highway speed to recharge a battery. In this case it is better to use a battery charger.
- You can actually warm up a battery by increasing consumption, but it will never be sufficient to ease starting the engine. Actually you might just use the last bit of power that would have started the car.

- There is no memory effect in lead-acid batteries. That is why they work so well in cars. If they lose capacity it is due to aging cells or bad maintenance.
- The applications in the car will only use as much of the capacity as they need. A higher capacity does not do any harm.
- When completely unloaded a car battery can reverse its polarity during the initial recharge.
- It is normal for a car battery to discharge at a rate of 1-25% per month.
- A weak or defective battery can affect the charging and the starting. If you replace the battery, alternator, voltage regulator or power, make sure all components are in good condition.

The truth about car batteries is: "Preventive maintenance of a car battery is very easy and should be done once a month in warm periods and whenever you change the oil for engine during periods of low temperatures".

CAR MAINTENANCE– *The Suspension System!*

The suspension system in a car is often considered to be one of the most critical systems in the car. This makes a lot of sense, as the suspension is predominantly involved in making sure that the car is stable and comfortable when driving. This means that if you have a car of any kind, it would be a good idea to try to make sure that the car's suspension is always in good shape. There are times when the system may develop problems that are hard to know of, especially if you spend most of your driving life on urban roads and do not drive fast.

The only way to make sure that such problems never exist is by making sure that you service your suspension according to the schedule set forth by the manufacturer. When you read your car's user manual, you will find that the manufacturer will always provide suggested intervals of service or replacement of such parts. If you can not get hold of such information, you can simply contact the manufacturer and ask them to provide the information that you need. Most of them can be reached through email or by phone call, so this is something that should not present much of a problem to you.

In many cases, you will find that it costs a lot to do major suspension changes to your car. Fortunately, there are many ways round this. You could decide to buy the spare parts you need online instead of from a brick and mortar store. There are many online stores that usually stock such items, and most of them are usually very cheap for various reasons. The only thing you would need to do is make sure that you buy products that are of high quality. This can be done by always making sure that you only buy from high quality stores.

In addition to that, you can also replace the parts on your own. For instance, if you own a Dodge, you can go online and buy a Dodge service manual. You can then use this to figure out what kind of tools and spare parts you need, and then go about the process of repairing the suspension on your own. This way, you are going to save on the cost of labor, and you can also ensure that you do a thorough job. This is just one of the ways of making sure that your car's suspension is always in good condition.

CAR MAINTENANCE - *Most Common Car Problems!*

Vehicles are no different than any other appliance we own, they will develop problems that need to be checked into. There are various ones that are considered the most frequent problems. However, we have list of them that should be checked to assure you of peak performance from your car.

Keep in mind, an ounce of prevention is worth a pound of cure, so check-ups are the best way to keep your car parts from failing.

1. *Brake Systems* - Brakes will wear out and replacement is a necessity. If there are noises such as squeaking or they feel different than they did, it is best to have them checked. The whole system should be checked. Most of the time it is only the pads that need replaced.

2. Lube-Oil-Filter - This is something that simply must be done. The recommendation from most all manufacturers is for the lube and oil and filter change every 3,000 miles you add to your vehicle. If you do not do this, you risk ruining the engine. Do not risk burning up the engine of your car because you did not have a simple oil change performed.

3. Cooling and Heating Systems - In the spring and fall are the best times to have your car in the shop for a check-up. This is a typical time for car repair shops to have discounts available for routine check up including your radiator.

4. Ignition and Electrical Controls - Your car's electrical system are prone to wearing out and a check up can prevent requirement of major repairs that will be expensive. Your car's ignition could fail or the controls that make the car run properly could be corroded or damaged. Checking these parts is very important.

5. *Steering and Suspension* - The steering is one part you can tell is not right. This may just be that the power steering fluid is low. Suspension is another important factor to controlling the car so these are two areas that need to have regular check-ups.

6. *Carburetor and Fuel System* - Has your gas mileage gone down? If so, it could be the carburetor or fuel system causing problems. These should be regularly maintained.

7. *Electrical* - The electrical system is a huge part of your vehicle. These parts all run and operate with the computer system in your car.These parts should be checked with a good electrical line specialist. So make sure the shop your car goes to for repairs has one.

8. *Transmission/Clutch/Rear Axle* - If your transmission is not operating properly, this can cause the clutch to burn out of your vehicle. Also affecting the rear axle, this is an extremely important part to check. An examination of this part of your vehicle is needed to make sure it is operating properly.

9. Air-Conditioning - The air-conditioning system works on fluids in order to work properly. This is extremely important in the summer months. Other parts including the hoses should be checked for optimum performance.

10. Exhaust System - If there is a black cloud of smoke chasing you down the road, the exhaust system needs a serious check-up. This keeps your car from running properly as well as adding to air pollution.

Vehicle check-ups are important to make sure your vehicle is running the best and that it will perform well for you for many years. Remember this when your check-ups are due. A 10 or 15 point inspection is the best so that all the parts can be checked. These are often put on sale at regular times.

CAR MAINTENANCE –*Some usefulTips!*

The truth about cars is that they do eventually start to have problems, no matter how much we may try to prevent it. Car problems are a loathed hassle to have to deal with. While there is no way to protect yourself from all car issues, there are steps that you can take to avoid them. Here are car maintenance tips that you can follow in order to prevent future troubles :

1. Engine Cooling System : An engine overheating will cause serious damage to your car. This is definitely something that you want to avoid at all costs. Make sure you regularly check your coolant and make sure that it is filled to the line. If you discover that you have a coolant leak, bring your car to be fixed right away.

2. Air Filter : Dirty air filters cause a loss of engine power and reduced gas mileage. Air filters need to be replaced periodically. You can check your owner's manual for replacement details for your car.

3. Spark Plugs and Timing Belts : Just as with air filters, these parts also need to be changed regularly to maintain engine efficiency.

4. Brakes : Having brake pads changed is an important money-saving maintenance tip. Brake pads are relatively inexpensive, but if you let the brakes bind it will be a costly repair.

5. Battery : Check battery terminals to see if they are corroded. Corroded battery terminals can cause lots of problems including low charge, trouble starting, and dim headlights. Also if you see any battery leaks, change the battery right away.

6. Oil : Oil should be checked at least once a month. If oil levels are low, re-fill to the line and check for leaks. Oil should be changed every 3 months or 3,000 miles, whichever comes first. Old oil can cause major damage to the engine.

7. *Tires* : Check your tire pressure at least once a month. The proper pressure should be listed in your owner's manual. Tires should be rotated at every other oil change to ensure that they wear evenly. There is a safe limit of tread wear and once that limit is passed, the tires are unsafe. Replace them.

8. *Windshield Wipers* : Old wipers can damage the windshield and replacing them is only a few amount. Besides damaging your windshield, they can be hazardous to your driving since you can not see out of your windshield if the wipers do not work!

9. *Clutch* : If you have a manual transmission, do not keep your foot on the clutch while driving. If you do this, you can cause serious transmission problems.

10. *Regular Service* : Taking your car to a professional mechanic will save you lots of money in the long run. Regular checks allow minor problems to found and fixed, preventing them from turning into major, expensive issues.

CAR MAINTENANCE -*Some Essentials!*

A car like any other machine is made up of hundreds of car parts that work together to make a car run. All of these parts are crucial to the proper working and smooth running of a car. It may be as trivial as a screw or nut in the car engine or car gearbox. A loose nut or screw can not only hamper the smooth running of the car, it may even prevent the car from running at all. If it is a crucial nut or screw, such as the ones used in the wheels of a car, it becomes more essential that they be properly fixed. The quality, servicing, maintenance, and or replacement of car parts over a period of time is essential and crucial for the proper running and long life of the car.

Timely and proper servicing and maintenance of a car and car parts ensures a longer life for the car with problem free driving. Properly servicing and maintaining the car, with timely replacement of parts with quality spare parts, ensures that the car does not ditch you when you need it the most, especially in an emergency. One of the most crucial parts of a car,besides the brakes and car engine, is the car gearbox system. A serviced and maintained car gearbox ensures a smooth ride.The opposite of this, a poorly maintained and serviced car and gearbox could be a nightmare ride, with loud noises and bumpy rides. An improperly working car gearbox could sometimes seem like a giant gnashing his teeth in rage. If you ignore the car gearbox and not attend to it in time, you may very well have to replace the whole gearbox in order to use the car.

Quality car parts, timely servicing and maintenance, and or replacing parts with good quality spare parts is not only essential, but crucial to the smooth running and long life of a car. If you can think of a car as a living, breathing object, then you can understand that just like any other animate object in this world, a car also requires food (oil and gas), servicing and maintenance (washing, oil changes, etc), or some tuning and adjustments of parts (heart surgeries, liver, lung, and stomach ailments) and sometimes replacement of parts (liver transplants, heart transplants). This just shows that if you take care of a car like a family member, it will provide trouble free service for years and be with you like a friend for life.

CAR MAINTENANCE – *Its Importance!*

Car maintenance is usually not one of our favourite activities in life. Most of the time, when using our cars, we take for granted that they are in a proper driving condition and sometimes can forget the importance of keeping them well maintained. Cars require general and routine maintenance over time and emergency repairs may be needed if you have an accident or if something goes wrong when you are driving. General car maintenance can help to prevent the latter by ensuring that your car is in optimal driving condition at all times. A well maintained car may also mean you are less likely to claim on your car insurance, potentially resulting in lower premiums over time.

Among the general car maintenance tips, getting regular oil changes at the appropriate period of time or usage is at the top of the list of importance. Many people forget when the time comes to get an oil change. Regular oil changes can help ensure that your car runs efficiently and that your engine is well maintained. Old oil or not enough oil in your vehicle can lead to overheating of the engine or other major engine and motor problems that can be expensive to fix. It can also affect air flow and other operational areas that make your driving experience pleasant and safe.

Tyres - Even people that do remember to get regular oil changes might forget to periodically rotate tyres. It is a good idea to rotate your tyres consistently. Most car care experts suggest you get a complete tyre rotation at least once every couple times that you change your oil. Tyre rotations can help you prevent your tyres from developing uneven wearing of the treads which can cause your car to lean in one direction or another. Your tyres are the foundation of your car's base and its connection to the road, so worn or improperly maintained tyres could be a big safety problem.

Another often neglected and very simple car maintenance tip is checking and filling the air in your tyres. Tyre pressure goes down over time just from natural use, but it can go down more quickly if there is a leak or deficiency. It is important to have the right amount of pressure and to have an even amount of pressure in your tyres. Having too low pressure wears on your tyre as it is stuck between the road and the rims of your car. This can lead to expedited wear and ruining of the tread. Having your car tyres overfilled (not using a gauge to fill) with air can cause dangerous blowouts that could prove fatal.

Tune ups - Regular tune ups and check ups also help ensure that you are not caught off guard with a dangerous car problem and a major expense. Still, even with all of the proper care measures that we have talked about, you are still at risk of a potential issue or accident every time you get behind the wheel. This is why having adequate car insurance is so important. Whether it is protecting your car's value or the health of you and others (liability insurance), you need to be protected before you drive.

Always remember these car maintenance tips:

- *Get oil changes.*
- *Look after your tyres.*
- *Get check ups and tune ups.*
- *Adequate car insurance.*

The Importance of Doing Scheduled Maintenance to Your Car

If you own a car, it is your responsibility to take care of it which means a lot more than just driving safely. In order to keep a car in working order, it is important to follow the suggested maintenance schedule for the vehicle. If you are good with cars you may be able to complete some maintenance items on your own, such as oil changes, filter changes, fluid top offs, and a tire pressure check. However, more complex work like brake maintenance, tire alignment, and belt checks are better left to the professionals. There are many reasons to keep up with scheduled car maintenance.

Keeping up with scheduled car maintenance can save you money. Sure, it may cost some money to bring your car in to a professional to get it done but it is better to spend some money up front than end up paying a lot more money later on. Regular maintenance lowers the chances that you will have to pay for costly repairs down the road, or even that you may need to get a new car altogether.

In addition to saving money, keeping up with a car's maintenance schedule can increase the life of your vehicle by several years. Cars are expensive and most people want to keep their car in working condition as long as possible. If you neglect to keep up with the maintenance schedule, the car will begin to deteriorate over time. Maintenance keeps small problems from becoming big problems.

By properly maintaining your car you are doing your part to avoid a break down and maintain the safety of yourself and your passengers. There is nothing worse than a break down or an accident that could have been prevented. Important safety features can wear down just like any other part of the car as it ages.

Keeping your car well maintained also preserves the resale value. If you like to sell your car every few years in order to upgrade, it is important to keep your car in the best shape possible. Keep all documentation and proof of completed maintenance in order to sell a car at a higher rate.

Every vehicle has a different maintenance schedule so it is important to check your owner's manual and stick to the schedule as best as you can in order to keep your car running smoothly for as long as possible. It will save you time, money, and a lot of headaches by doing so.

CAR MAINTENANCE– *Regular Practices!*

Regular safety checks are a great way to avoid unnecessary car trouble. In a lot of ways, regular maintenance is the best kind of insurance for your car. Poorly maintained vehicles can lead to accidents which often result in higher car insurance quotes. Not only will your car have a longer lifetime and be safer to drive, but this is also the simplest way to avoid unnecessary fines and collisions. Regular maintenance does not mean spending hundreds of dollars and major tune-ups every year, but rather smaller efforts along the way. Small efforts to maintain the different systems in your car will make it less likely that you will need a major tune-up.

Take care of your car and it will take care of you. Just like the majority of mechanical items, regular maintenance is needed to keep a car running smoothly and safely. If a vehicle is well cared for and regularly maintained, the car has a drastically lower risk of a mechanical failure resulting in a car accident.

Here Are The Basics Regular Practices.

Tires: Roughly 20,000 crashes per year are caused by tire blowouts and failures. Maintaining proper inflation is one the simplest ways to avoid meaningless crashes. Get a pressure gauge and check your tire pressure often - once a month should be ade uate. It is important to also check the depth and wear of your tires. When tires are worn down they have less traction and it becomes more difficult to stop suddenly. A tire's tread will show whether it has proper traction and if the wheel alignment is on track.

Brakes: Regular checkups will insure that your brake pads and rotors are in good condition. Brakes are the first line of defense in a dangerous driving situation and need to be kept up. Old brake pads can drastically slow the time that it takes you to stop and make it more likely that you will crash in an emergency stopping situation.

Fluids: Many of a car's most crucial systems depend on whether or not there are appropriate levels of fluids. Check your car once a month for leaks and tend to any if you find them. Regular oil changes are the easiest way to insure longevity in the life and function of your vehicle, let along better operational safety.

Belts and Hoses: Examine under the hood and check for any loose connections, leaks, or cracks and tend to any issues that you discover.

Lights: Make sure that all your lights work so that other drivers can see you car at night and know if you plan to turn, reverse, etc.

Summary

By performing the following "spot checks" every month, you can keep your car in good working order:

Every Month - Oil level, hoses, belts, tire pressure, coolant, air filter.

Every 3 Months - Oil change, oil filter, windshield washer fluid, brake fluid, transmission fluid, battery life, battery cables.

Every 6 Months - Wiper blades, all exterior lights, horn, breaks, spare tire, exhaust system, shock absorbers.

It is important to check the particular needs of a vehicle. A detailed checklist of regular car maintenance for most vehicles can be found online. A mechanic will also provide you with information on what services they performed and when they performed them, making it easier to stay on top of this important checklist.

Adhering to this maintenance list will have two major benefits. First, it will save money. People who do not change their oil frequently are being financially irresponsible. A fast, cheap oil change is much less expensive than having to purchase a new engine.

The second benefit is it may prevent an accident. Brakes should be checked every 3 months. If a person fails to do this, their brakes may unexpectedly stop working.

Not only is this putting a driver at risk of injury, it also endangers the lives of all others around them. Financially, this may take a toll. The irresponsible car owner will have to pay their deductible.They will consequently see a rise in their insurance rates.

CAR MAINTENANCE - *Carpets and Upholstery!*

Soil particles, dirt and dust acquired by the upholstery and carpets inside the car could be easily eliminated by following some useful Upholstery tips. This book reveals several techniques and tools that could help in efficient car upholstery cleaning. Unmoving upholstery fixed in the cars could be a difficult process for many of the car owners. Daily usage of the car could easily allow pollens, dust, soil, and other unwanted things to ruin the carpets and get attached in the upholstery materials or fabrics.

In addition, the general habit of having snacks or food inside the car could lead to gathering of marks and stains. Here are some of the tips to clean carpets and upholstery inside your vehicle and to maintain their durability as well.

- *Vacuum Cleaning*

One of the most important methods of cleaning the interior of a car is by using car vacuum cleaners. But, if you want to have a clean vacuum process, you can hire any of the expert services for the same. The cleaning experts are capable of removing the front seats of the car and vacuum each and every corner of the car thoroughly.

- *Removing Stains*

The choice of dirt removing solutions depends on the kind of material the upholstery is made up of. For example, the cloth-made car seats could be simply cleaned using water and soap solution. But, for vinyl and leather seats, one needs special cleaning solutions that are available in the market. You could learn about the right cleaning solutions from a professional in the field.

There are several cleaning services providers available in the market. You have to choose the one which suits all your requirements and budget. Select the one which provides good service at reasonable price.

CAR MAINTENANCE - *Installing and Maintaining Car Speakers!*

Installation forms the most important part of the car speakers. Various cars have various specifications as per the sound set up and installation process. This small crash course on the step by step procedure will act as a basic car speakers guide for you. The first thing you should do before starting installation is to become acquainted with a few essential tools like screwdrivers.

You should know that each vehicle has its own specifications related to the installation issues, which vary with factory speaker location, the depth and height for mounting. Another factor to be taken into consideration is the car factory wiring. However, the most important thing required is your time and effort.

For installing car speakers, you need a few varieties of screwdrivers along with torx drivers and drill bits. Do not miss out on the socket wrench set along with the wire cutter and stripper tool. You may even require soldering iron and solder crimping tools as well as connectors along with panel removal tools and a clip remover.

However, if a clip remover is not available you can always opt for a screwdriver covered with a shop rag. You may also require a file, electrical tape and a knife. You will be required to use the wiring harness to attach the new speaker, or solder or crimp the connections, as the installation may require.

A complete component speaker system includes separate woofers, tweeters and crossovers that provide a surround sound while in the car.

A few things that you should remember, and that will help you in the long run are -

Each and every vehicle is different, so you may have to deal with them differently. The first basic step lies in identifying your vehicle. Once you have grown comfortable with your car, identify the location where you want to mount your speakers. Once you have selected the location, choose the size. Choose a speaker that comfortably fits in your car.

Make use of the factory grills and brackets while fitting in your speaker. Speakers labeled E Z will fit in the factory speaker openings. Try to use the magnet to fit the available space and do not allow the tweeters to interfere with the grills.

The speakers labeled P fit with the help of mounting brackets, which come free with the purchase of the speakers. If that does not help, you can go ahead and do a few alterations like cutting a new screw hole, cutting the required area of the pressboard or the metal to make room for mounting the speakers that are larger than the factory speakers.

In a setting of component speakers, the component woofers can be installed in the location where the factory speakers were located. If you have a set of tweeters too; you may have to think of a place where you can install them. The tweeters require custom installation. Thus to do that, it involves drilling holes either in the door panels or on the dashboard. If it is an area like the dashboard, use your tools carefully to avoid scratches on the surface.

CAR MAINTENANCE – *Washing With Modern Equipments!*

If one requires a well maintained car it should be washed at least once a week to keep it looking bright and shiny. No one would be in a position to take one's car to the washing centre each week so a set of car wash equipments should be maintained in the home. The equipment can be very simple tools that are light and versatile and easy to store.

For instance if one has a small car the washing equipment can be a wash mitt. One can easily slip the hand through this and with water and the required soap or detergent one can give the car a thorough wash. If the car is a big one, like a truck the debris would be hard on the surface and would require a brush as a washing tool. The brush should be of soft bristles as the hard bristles would leave scratches on the painted surface of the vehicle.

Brush sizes and styles vary depending on the part of the car that has to be cleaned. For instance one can use a bug and tar together with a soft brush. It would provide the best results without leaving any scratches on the paint. But one should always remember that all car wash equipments are not hundred percent effective unless they are used properly.

But to make it more effective the self service car-wash pump serves as useful cleaning equipment. The other useful equipments include hose clamps, piston pumps, surface cleaners, air compressors, vacuum cleaners, claws and cuffs, steam and cold water mixing units etc.

The choice of polish and wax is mandatory for the perfect finish of the car. Car wash can be done regularly but polishing is advisable once in six months or one year since frequent polishing could spoil the outer paint completely. Wax is of a softer nature and it has the conditions of giving the car a super shine while filling in small dents left in the paint due to regular use. Polishing should be done using a buffer that will remove stubborn stains and dull finishes left on the paint's surface. The wax adds luster and long lasting shine to the car.

These two car wash equipments are not hard to acquire. Earlier people used to admire the look of cars in the exhibitions and long to have one like that. Now with the introduction and mass production of quality wax car owners can enjoy that luxury of a sparkling car at an affordable price.

Steps in Maintaining Your Car Paint

Very few people know how to maintain their car's paint. Almost everyone will always just wash and dry their car. Occasionally when they feel that they need it to look good they will have it waxed.

- *Washing*

Washing is what we are most familiar with. Because washing is intuitive we think that the main way of protecting our cars is through washing.

Although this can be somewhat true, washing actually removes the clear coat for cars that have it. This does not happen in one washing but through numerous washing sessions especially with detergents not meant for car use.

- *Clay Bar*

Claying a car would be difficult for anyone unfamiliar with car detailing to put together. Indeed much confusion happens when a Clay Bar should be used.

Clay Bars are meant to be used after washing. They work by clearing the paint of imperfections. Contaminants that were not taken out by washing can be effectively removed by a clay bar.

- *Polishing*

Polishing may sound like the last step in maintaining your paint but actually it is not. People might mistake polishing as a process of making the car glimmer or shine.

When we say polishing here it actually is more like cleaning. A polish contains very light abrasive materials that dig into the paint to remove scratches.

Polishing is also responsible for removing those fine cobweb shaped swirl scratches that ruins the finish of a car.

- *Waxing*

Waxing is a protective process. A wax creates a layer of protective material that acts as a shield. You can think of this as an extension skin for your car. This is probably the most important step as it prevents the elements like the sun and weather from ruining your paint.

Waxes generally come in different forms. The most popular among them is the carnauba wax coming from part of South America. Newer synthetic "waxes" are not even waxes at all. But they do have the same bonding effect that carnauba based waxes do.

This steps discussed in maintaining your car is not a rigid system. Waxing your car for example can be done every 3 months or so. While washing it down with a car shampoo should be done more frequently.

If however you want the maximum treatment to maintaining your car's paint then doing this steps twice or thrice a year will ensure a car that will hold on to its shine for a long time.

Things You Should Avoid When Washing Your Car

There are a few secrets to washing your car that will not damage your paint. This book will discuss some of the things that you should avoid when washing your car to prolong the life of your car paint.

- *Do not wash the car in direct sunlight*

Perhaps the greatest mistake you can make when washing cars. Washing the cars on direct sunlight can have a dangerous effect on your paint. Each water droplet will act as a magnifying glass.

This produces little laser like beams that will create microscopic pits on your car paint. The end result is a premature chipping of car's paint especially in the front portion.

- ***Do not use the same mitts for upper and lower parts of the car***

If you use mitts then it is best to have two of them. One for the upper half of the car and one for the lower half is a good way to ensure that dirt on the lower portion will not contaminate the more visible upper part. A lot of cars have a line that runs through the level of the headlight that will help you distinguish from the lower or higher parts.

You can also look at the sides of your car when it is dirty.You will notice that from the bottom of the car up to a certain point will be abrasive. Or even contain grains of asphalt.Take great care that these grains do not make their way up to the upper half of the car.As they can cause deep and visible scratches on your car's finish.

Using two mitts or car washing towels can help you with this.

- ***Do not allow your car to dry in air***

Allowing you car to dry without your intervention can leave nasty water marks on your paint. This is caused by tap water with higher mineral content. They can be nasty and be very difficult to remove.

No doubt you have seen cars sprinkled with so much of these little white circles.

- ***Do not forget to wash your wheels***

Wheels and tires are among the dirtiest part of your car. A lot of us neglect to wash them because they will get dirty again as soon as we drive. However if you leave them unwashed for a period of time they can be very difficult to remove.

They will accumulate into a greasy paste like layer on your wheels. This can take hours to scrub off. To avoid this wash your car's wheel regularly with the help of wheel detailing products like brushes.

If you learn to avoid the things mentioned above your car will reward you with a finish that will be shiny and free of damage for a long time. This might even help you earn extra money when it is time to sell your car. So make sure to take care of your car's finish by being well informed.

CAR MAINTENANCE -*Seat Covers For Protection!*

There is nothing like driving a car that does not only get you to your destination, but also feeds your senses with comfort and beauty. Even when you are merely a passenger, you would always want to enjoy each moment and view that you happen to be passing by.

When your car's interior is in good shape, there is no question as to relaxing feeling that you are able to relish while you are on the way to another long day at work. If you are going on a road trip, having a well-maintained car interior can let you enjoy all that freedom on the road more than if you were on a seat with foam or springs sticking out. If you value the time you spend in your car, it would be great to take care of its interior.

Car seats are particularly powerful when it comes to giving that look and feel inside your car. That is why you would always want to keep it looking and feeling good. And one way to do this is to get car seat covers. Seat covers will notjust look good, but they will also to function to protect your original seat covers so they maintain their good form for a longer time. Everyday wear and tear can leave those original covers with tears, scratches and even pen marks when you have kids riding with you. To protect your seats from all these damaging elements, you can make use of seat covers. By using it, it can give you more advantages in the future.

There is a great variety of materials used for seat covers, from leather to cotton, and it all depends on your personal preferences. You can buy ready-made covers or you can have yours custom-made so you can get that perfect fit and comfort on your car seats. Should you go for ready-made items, you can look for those that are created for the exact brand and model of your car. Yes, you can make your order simply by giving your car's details, and then they will send you good-fitting covers for those seats. Again, if you would like something that follows the exact contours of your car seats, have those covers custom-made. Seat covers will be very easy on the budget in terms of cost and the possibility that you could just be spending more getting those damaged seats fixed at a car upholstery shop.

CAR MAINTENANCE -*Importance of Car Covers!*

Nowadays, cars are very important especially when carrying out daily activities. Some people buy for luxury. For whichever reason that you are buying it is important to protect it in order to maintain its value. Protecting your vehicle should not be seen as a waste of time. It is important to have a cover that will be adequate protection.

Car covers are important for protecting your vehicle. Due to the present nature of economy it is important to have a cover that is strong enough. Quality covers should be able to protect your vehicle for a very long duration.Many people value their car as a great asset thus they can go to greater length to make sure that their car looks presentable. It is important to keep your car in the best condition possible. Cars require to be properly maintained. You should have a good cover for your car type.

When you are not using your car it is important to protect it using car covers. They help in maintaining the aesthetic value of your car. They give protection to your car from external factors. If a person writes on your car using a sharp object, you will incur cost while trying to repair the damage done however you can reduce these maintenance costs by using covers. Some people will scratch or damage your car willingly or unwillingly; however, the cover will prevent such cases occurring. Some people view these covers as an expense but they provide more benefits. For those who know the benefits of covers, they never fail to use them.

When you park your car for a long duration, it is important to cover it. Even if you have parked your car in a private car park, it is important to cover it as other cars will produce smoke that contains harmful products. When you have not covered your car, it will become dirty due to the smoke.

During a bright sunny day, there are usually ultra violet rays which are produced by the sun. Ultra violet rays are harmful to the car as they cause the car's paint to appear dull. This will make your shiny beautiful car not to be attractive; to prevent this, you should use a car cover. When it is hot, the car absorbs heat and when people get inside, they become uncomfortable due to the hot temperatures inside. When there are extreme high temperatures inside your car, they will cause damage to your car seats' covers. All these disadvantages brought by high temperatures can be reduced by using the covers. The covers will help in keeping your car cool even when it is hot.

When your car is not in use, dust usually accumulate on it. Some dirt particles accumulate in the car's internal parts thus increasing its maintenance costs. It is important to reduce these maintenance costs by proper use of car covers. It is important to know how to use the cover properly so that you can realize the mentioned benefits.

CAR MAINTENANCE – *Keeping It New!*

Every one who buys a new car intends to keep it that way. But cars, among other things, wear out owing to daily use, pollution, traffic and other factors. This is why maintenance of the car, especially its engine is very important.

To keep your car looking as good as new, you must wash it every weekend, and scrub it every day. But more than the exterior, it is what inside the car that matters most. Car engines are the most crucial part of the vehicle. The engine is as important as the heart is to our body. Therefore, it matters a lot how you treat your car and what you 'feed' it, for its proper functioning.

Car engine maintenance requires that the oil is changed every 3000-5000 miles. This will ensure your car gives you better mileage, even if it is 5 years old. Synthetic oil is known to be a good lubricator for the engine as it breaks down slowly. There are many oils that use synthetic technology such as Castrol, Mobil and others.

Get your car serviced regularly, from authorized service centres. Make sure you get your oil filter changed periodically, but do not change the type of oil filter. Also keep a check on the oil levels. It should be bent towards 'F' or full. Do not overfill it. If the oil level remains low, it might be because of a potential oil leak in the car engine area. It should be reasonably damp but not too wet. Also check to see if your car is leaving a trail of oil. If it is,rush it to your service centre for repair.

Do a dipstick oil test regularly. Once you put the dipstick in the oil filter, notice the color. Well maintained car engines have a golden, honey-colored tone. If it is black, it is a sign of carbon deposits on the filter. It means your car consumes more oil than required and needs to be serviced.

Car engine maintenance requires you to listen to your car - literally. When you start your car, listen to the rev of the engine. It should not rattle or whistle. The level of noise indicates the level of depreciation of your car engine. The faster you catch an issue with your car, the faster you will be able to solve.

WAYS TO AVOID CAR ACCIDENTS – *Extra Tips!*

With the increasing number of cars on the road, the number of accidents is also increasing. These accidents are mostly caused due to the negligence of people and could have been avoided if they become little conscious while driving. Many of them have proved fatal.

Here are some advices that will help you to prevent accidents.

- *Driving while drunk*

Most of the time person does not know what he is up to while drunk. Avoid driving if you are drunk because you are not in your senses and there are greater chances of an accident.

- *Felling sleepy*

Do not drive if you are tired and are feeling sleepy. It is advisable that you park your car for a while and take rest for a short time.

- *Over correcting*

People try to dodge the object they see on the road and do not want to hit. While dodging they lose control and results in an accidents. The object may be weed, road cones, etc. It is better to hit that object because it will not cause much damage as an accident would.

- *Stopping on red signal*

It is strictly advisable to stop on red light even if you are in hurry.

- *Turning left on yellow light*

Avoid turning left if the signal has turned yellow. Most of the people who want to go straight turn left when they see that the signal has turned yellow. This may prove fatal because the cars that are already waiting for turning left may not see you and you may get hit by any one of them. Also start slowing down your car if you see yellow light.

- **Maintaining less distance**

Some people do not maintain proper distance and follow the other car too closely. This blinds the driver of the car that is following too closely and may hit the car ahead if it applies brakes due to some reason. It is essential to maintain some distance because the braking system takes some time in making the car stop. Many accidents can be avoided just by maintaining a distance.

- **Parking lane**

The side of the road is designated for parking your car. Avoid driving in the lane right of white lane because they are for parking purpose and your little negligence can result in an accident.

- **Not adjusting your speed according to the conditions of the road**

Adjust your driving speed according to the circumstances. Follow the speed limits that are mentioned on the side of the roads. Driving too fast at a place where you need to slow down may result in accident. Maintain your speed that is less than the maximum limit or that you can handle easily.

- **Losing concentration while driving**

When driving, do not pay attention to any other thing drive carefully. Most people do not concentrate while driving and when some thing happens they are unable to react properly. This leads to an accident. Avoid using cell phones, eating, etc when you are at the driving seat.

- ***Carefully change the lanes***

First make sure that the lanes are clear and then change your lane. Look for the approaching cars and make your decision accordingly.

- ***Strictly follow traffic rules***

Traffic rules are for your safety. Follow them to ensure your safety. Avoid turning from the place where it is not allowed.

Follow all the above rules to ensure your and your family safety!

CONCLUSION

Maintaining your car can help make your car last longer, help you save money on gas, help save you money on unnecessary repairs, and so much more. It is not difficult to do on a regular basis, nor is it hard to learn. You do not need to be an auto mechanic to know how to properly maintain your car. This book has provided you with the necessary steps to take to keep your car running smoothly, and help you learn how to do each one.**THANK YOU!!!**

40976156R00080

Made in the USA
Middletown, DE
02 April 2019